Synergistic Soil Microbiomes: Harnessing Rhizobia and PGPR for Sustainable Legume Production

Co-Inoculation Strategies to Enhance Nodulation, Nitrogen Fixation, and Biomass Yield in Chickpea

Andargachew Belay Abrham
Selam Mesfin

Synergistic Soil Microbiomes: Harnessing Rhizobia and PGPR for Sustainable Legume Production

Co-Inoculation Strategies to Enhance Nodulation, Nitrogen Fixation, and Biomass Yield in Chickpea

ELIVA PRESS

ISBN: 978-99993-4-503-3

Cover Design: Eliva Press
Cover Image: www.ingimage.com
Email: info@elivapress.com
Website: www.elivapress.com

This Eliva Press imprint is published by the registered company Eliva Press Global Ltd. part of Eliva Press S.R.L. Publishing Group

The registered company address is: Pope Hennessy Street Level 2, Hennessy Tower Port Louis, Mauritius
Eliva Press S.R.L. Publishing Group legal address is: Bd. Cuza-Voda 1/4 of. 21, Chisinau, Moldova, Europe

Synergistic Soil Microbiomes: Harnessing Rhizobia and PGPR for Sustainable Legume Production

Subtitle:
Co-Inoculation Strategies to Enhance Nodulation, Nitrogen Fixation, and Biomass Yield in Chickpea (Cicer arietinum L.) under Greenhouse Conditions

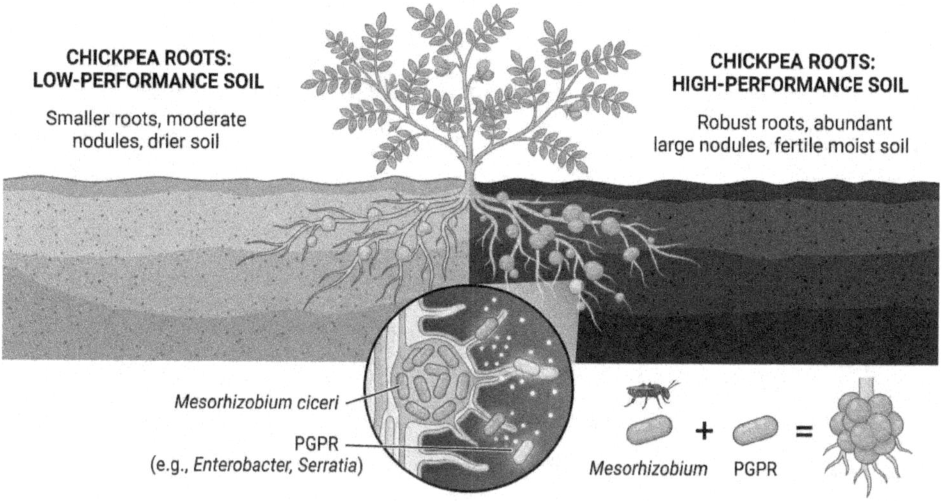

By

Andargachew Belay Abrham[1] and Selam Mesfin Ayalew[2]

[1]Department of Biology, College of Natural and Computational Sciences, Injibara University, Injibara, Ethiopia

[2]Department of Plant Sciences, College of Agriculture and Environmental Sciences, Bahir Dar University, Bahir Dar, Ethiopia

ABSTRACT

This document discuses about the chickpea production constraints in Ethiopia and the study conducted on evaluating the effect of single and co-inoculation of rhizobia and PGPR isolates on the symbiotic effectiveness of chickpea under greenhouse conditions for improving the productivity of chickpea. Three rhizobia isolates were chosen based on eco-physiological tolerance, antibiotic resistance, PGP, biocontrol properties and symbiotic effectiveness. The PGPR isolates were selected based on PGP, biocontrol properties, stress tolerance and antibiotic resistance abilities. In single inoculation, the three rhizobia isolates induced nodule numbers ranging from 37-46 /plant in Dembia soil and 35-42/plant in Adet soil. Co-inoculation treatments generally showed an average increase in nodule numbers by 21-125% compared to single inoculation treatments. The highest nodule dry weight, 301 and 237 mg/p was accumulated by plants inoculated with the consortium on Dembia and Adet soils, respectively. Regarding the shoot dry weight, in Dembia soil the highest shoot dry weight (4.323 g/p) was accumulated by plants inoculated with consortium, followed by 3.817 g/p and 3.536 g/p co-inoculated with GUCR-30 (Mesorhizobium sp. HKG230) + GUCRB21 (Enterobacter mori) and GUCR-19 (Mesorhizobium amorphae B19) + GUCRB76 (Serratia marcescens). Chickpea inoculated with consortium followed by GUCR-30 (Mesorhizobium sp. HKG230) + GUCRB21 (Enterobacter mori) in Dembia soil and GUCR-30 (Mesorhizobium sp. HKG230) + GUCRB76 (Serratia marcescens) in Adet soil displayed the highest shoot total nitrogen content. Co-inoculation of rhizobia and PGPR isolates led to a significant increase in nodule number, nodule dry weight, shoot dry weight and shoot total nitrogen compared to single inoculations and controls. A further field experiment is recommended for upgrading these isolates into chickpea inoculants.

Key words: Biological nitrogen fixation, Chickpea, Co-inoculation, Microbial consortium, Nodulation, Sustainable agriculture, Symbiotic effectiveness, Plant growth promoting rhizobacteria

Acknowledgement

We would like to acknowledge Injibara University and the Bahir Dar University for their kind assistance in providing the laboratory facilities, greenhouse and all the required consumables and equipment.

Dedication

This work is dedicated to my My sweet and loving kids, Meaziyos and Lael.

TABLE OF CONTENTS

Contents

1. INTRODUCTION

Agriculture is the fundamental driver for Ethiopia's economy and long-term food security as it contributes 40% to the GDP, employs some 79% of the population, accounts for 80% of foreign earnings, and is the major source of raw materials and capital for investment and market (MOA, 2023). According to FAO statistics, 25% of Ethiopia's population suffers from under nourishment, while 19.6% dwells in a situation of severe food insecurity (FAOSTAT, 2024). Therefore, there is a need to enhance agricultural productivity by improving access to resources and promoting sustainable practices like soil management.

Legumes are an integral part of the ecosystem that contributes to soil fertility and sustainable agricultural development across the globe and ensures nutritional security, particularly among the poor segment of society. Common bean, chickpea, cowpea, groundnut, lentil, pigeon pea, and soybean are amongst the seven major legumes are cultivated worldwide and highly preferred ones across the globe (Nigam *et al.,* 2021).

These legume crops are categorized as cool season legumes and tropical legumes where lentil, faba bean, field pea, chickpea, grass pea, lupin and fenugreek are cool season legumes where haricot bean, mung bean, cowpea, and soybean are warm and tropical legumes, respectively.

According to CSA (2022), legume crops covered 13.75% (about 1,676,888.30 hectares) of the grain crop area and 9.66% (about

31,680,193.33 quintals) of the grain production. Chickpea (*Cicer arietinum* L.) is considered a highly nutritious and economically accessible food (Sofi *et al.,* 2020).

India, Australia, Turkey, Ethiopia, Russia, Myanmar and Pakistan are some of the countries are the highest chickpea producers in the world. In Africa, Ethiopia is the first country in chickpea production and production area coverage (FAOSTAT, 2024).

Chickpea is cultivated after the rainy season due to their impressive ability to withstand low moisture stress, utilizing residual soil moisture. Furthermore, chickpea provides as source of supplementary dietary protein for individuals who cannot afford animal products and serves as a significant source of income for farmers through local and export markets. Lastly, the crop residue from chickpea is used as a valuable source of animal feed in the highland areas, where population pressure has resulted in a shortage of feed (Tibebu Belete *et al*., 2017; Setotaw Ferede *et al*., 2019). Besides its nutritional quality and source of income, chickpea plays a tremendous role in soil fertility by improving symbiotic N_2 fixation in association with rhizobia (Assefa Funga *et al*., 2016; Zehara Mohammed *et al*., 2020; Atsede Muleta *et al.,* 2021).

Even if chickpea is widely grown in Ethiopia, its productivity is still very low. It has been reported that the mean chickpea yield in Ethiopia in farmers' fields is below 2000 kg ha^{-1}, which is far below its potential yield of > 5000 kg ha^{-1} (Zewdie, 2018). The low productivity is mainly due to several biotic and abiotic production constraints. Consequently, chickpea production fluctuates annually with an erratic harvest determined by biotic and abiotic stresses (Aslam *et al.,* 2018).

1.1. Chickpea production constraints in Ethiopia

In Ethiopia, chickpea cultivation has increased by 23% from about 163,067 ha in 2003 to about 201,274 ha in 2022, while total production reached 396,510 tons, with average productivity of 1.97 tons per ha-1 (CSA, 2022). Despite its progress the productivity is below the crops potential. There are different factors affecting the productivity of chickpea. Different researchers reported chickpea production constraints in Ethiopia, some of the major constraints are discussed below.

❖ Soil fertility and nutrient deficiency

A 2024 field experiment in Southern Ethiopia (Cheha district) showed that low soil fertility particularly deficiencies in nitrogen (N), phosphorus (P), sulfur (S), boron (B), and low organic matter is a major constraint. This limits nodulation and biological nitrogen fixation in chickpea (Gashaw Nahusenay *et al*., 2024).

❖ Soil acidity/salinity

A very recent study (2025) evaluated 64 Ethiopian chickpea accessions under acidic soil conditions at two sites (Emdebir and Holetta). Significant genetic variation was found for acid-soil tolerance. Some genotypes performed consistently well (good yield + yield stability) across acidic sites, suggesting potential for breeding acid-tolerant chickpea cultivars. Another study identified aluminum (Al) tolerance in Ethiopian germplasm: Al toxicity (common in acidic soils) is a constraint, but some accessions show root-growth tolerance. A review argues that despite the prevalence of acidic soils (≈ 43% of Ethiopian arable land) and Al toxicity, Al tolerance has not been sufficiently prioritized in breeding. (Hawi Negusse *et al.*, 2025). Breeding for salinity-tolerant chickpea should also accelerated.

❖ **Disease**

✓ **Ascochyta blight**

A 2023 paper mapped the distribution of Ascochyta blight (*Didymella rabiei*) in central Ethiopia and analyzed factors driving epidemics. The disease remains a serious biotic constraint, contributing to yield losses (Shugute Addisu *et al.*, 2023).

✓ **Fusarium wilt**

Fusarium wilt is one of the most destructive diseases of chickpea (*Cicer arietinum* L.) in Ethiopia, causing significant yield losses across major chickpea-producing regions including Amhara, Oromia, and SNNP. Caused by soil-borne fungus: *Fusarium oxysporum* f. sp. *ciceris* (Foc), Produces long-lived chlamydospores, surviving in soil for >6 years. Favored by warm temperatures and dry stress conditions (Jimenez-Diaz *et al.*, 2015).

According to a review, the disease incidence (i.e., percent of plants affected) in surveyed regions ranges (e.g., in Amhara region, incidence ranged from ~11% to 58% depending on district). Fusarium wilt causes up to ~30% yield loss, though this varies with environmental conditions (Yilma Abebe *et al.*, 2025). Several field surveys show that fusarium wilt is widespread in central and northern Ethiopia, especially in: Shewa (East, North, West Shewa), South Gondar, South Wollo, Arsi and Bale zones, Gurage and Gojjam region (Tafesse Gezahegn *et al.*, 2020).

❖ **Climate Variability / Drought**

A 2025 study used an autoregressive distributed lag (ARDL) model to assess the impact of climate variability (rainfall, temperature) on chickpea production in the central highlands. The analysis showed that climate variability significantly influences chickpea yields in both short and long term (Tariku Tirfeessa *et al.*, 2025). Historically, drought (especially terminal drought) has been a key abiotic stress; while this is not entirely new, recent work continues to highlight climate variability as a key risk factor (ICRISAT, 2023).

❖ Agronomic / Crop Management Constraints

A 2024 validation experiment in the central highlands (Ada'a district) tested two drainage methods: broad-bed & furrow (BBF) vs. flat-bed (farmer practice), using both an improved chickpea variety (Natoli) and a landrace. Results showed some yield gains (e.g., 9.8% for the landrace with BBF), though not all differences were statistically significant. This suggests that water management (especially in vertisol soils) is a nontrivial constraint. (Assefa Gonfa *et al.*, 2024). According to ICRISAT (2023), a major constraint is technical inefficiency / low farmer knowledge. Their study (681 farm households) suggests that improving farmers' technical knowledge (e.g., better use of inputs, proper fertilizer, more effective pesticide use) could nearly double chickpea yields.

❖ Socio-economic / Institutional Constraints

A 2022 study in Gondar Zuria Woreda analyzed smallholders' constraints using Henry Garret's ranking technique. Key ranked constraints included: disease & pests, supply/cost of improved seed, and input cost / supply issues (Asrat Mulat *et al.,* 2022).

The usage of chemical fertilizers to enrich soil with nutrients in high-input cropping systems is often deemed a necessity to achieving optimal crop yields. However, their efficiency is hindered by factors, such as volatilization, denitrification and leaching (Fahde *et al.*, 2023). Prolonged use of chemical fertilizers can negatively impact soil ecology, harm the environment, degrade soil fertility, and harm human health (Pandey *et al.*,

2012). Pollution problems leading to public health hazards necessitated the development of technologies that are sustainable and eco-friendly, which could reduce the application of synthetic fertilizers (Zhang *et al.*, 2021). In sustainable agriculture, the application of beneficial microbiomes as biofertilizers has emerged as an innovative and environment-friendly technology for improving soil fertility and plant growth (Ullah *et al.*, 2020; Fasusi *et al.*, 2021).

Numerous agricultural systems depend on symbiotic relationships between leguminous plants and rhizobia. These rhizobia create root nodules on leguminous plants and transform atmospheric N_2 into a form that plants can use (Abd El-Azeem, 2022). Environmental factors such as low numbers of rhizobia in the soil, high temperature, salinity, low clay content, and increased concentrations of heavy metals and pH conditions that are harmful to the survival of rhizobia in the soil, frequently cause a threat to the production of legumes (Denton *et al.*, 2013).

Bacteria obtained from the rhizosphere that can produce and secrete metabolites, which promote plant growth after colonizing their roots are known as Plant growth-promoting rhizobacteria (PGPR) (Beneduzi *et al.*, 2012). PGPR includes members from various genera like *Agrobacterium, Arthrobacter, Alcaligenes, Azotobacter, Acinetobacter, Actinoplanes, Bacillus, Frankia, Pseudomonas, Rhizobium, Micrococcus, Streptomyces, Xanthomonas, Enterobacter, Cellulomonas, Serratia, Flavobacterium, Thiobacillus* etc. (Glick & Gamalaro, 2021; Kumar *et al.*, 2021). Different PGPR have been reported to promote plant growth and crop yield by increasing nutrient availability and uptake, producing plant hormones and suppressing soil-borne diseases (Santoyo *et al.*, 2021).

Currently, the combined inoculation of PGPR strains to improve the nodulation and nitrogen fixing potential of the rhizobial strains has received considerable attention. Upon inoculation, PGPR helps the plant to withstand drought stress (Ilyas *et al.*, 2020), salinity (Bharti *et al.*, 2016) and biotic stress (Verma *et al.*, 2016). PGPR Inoculation has been reported to enhance seed germination, soil fertility and plant growth by producing auxins, ethylene, gibberellins etc. (Jang *et al.*, 2017; Tahir *et al.*, 2017). Several studies (Mirza et al., 2007; Rajendran et al., 2008; Hungria et al., 2013; Sánchez *et al.*, 2014; Korir *et al.*, 2017; Laabas *et al.*, 2017; Adal, 2018; Sintie, 2018; Kumari *et al.*, 2020) have demonstrated that co-inoculation of PGPR and rhizobia enhance nodulation, nitrogen fixation, and yield of several legumes including chickpea.

In Ethiopia, different research works have been conducted on the diversity and symbiotic properties of rhizobia of legume crops. However, there is no comprehensive research conducted on the integrated application of rhizobium and PGPR on chickpea except for a screening study on phosphate solubilizing bacteria from the rhizosphere of chickpea by Midekssa *et al.* (2016). Hence, this study aimed to evaluate the effect of single and co-inoculation of rhizobia and PGPR isolates on the growth and symbiotic effectiveness of chickpea on soil culture under greenhouse conditions.

2. MATERIALS AND METHODS

2.1. Sources of Rhizobia and PGPR

Rhizobia and PGPR isolates were previously collected from the nodules and rhizospheric soils of chickpea from the central and south Gondar zones of the Amhara region, Ethiopia. The three rhizobia (GUCR 19, 30 and 55) and four PGPR (GUCRB 4, 21, 76 and 124) isolates genetically identified into the genus *Mesorhizobium* spp. and *Alcaligenes* sp. (GUCRB 4), *Enterobacter mori* (GUCRB21), *Serratia marcescens* (GUCRB76) and *Brevibacillus brevis* (GUCRB124), respectively deposited in culture collection centers of biology department at Injibara University. The isolates were used as sources of inoculants. The rhizobia isolates were chosen based on eco-physiological tolerance, utilizing different carbon and nitrogen sources, intrinsic antibiotic resistance (IAR), plant growth promoting properties (PGP), biocontrol properties and symbiotic effectiveness (Table 1). The PGPR isolates were selected based on their PGP, biocontrol properties, stress tolerance and IAR abilities (Table 2).

2.2. Single and Co-inoculation experiments in different soil conditions under greenhouse condition

The effect of single and co-inoculation of rhizobia and PGPR isolates on the symbiotic effectiveness of chickpea was studied in pot experiments under greenhouse conditions, using two different soils. Two soil samples were collected from the districts of Adet (11°28'42.6"N, 37°48'07.8"E), which had never been inoculated previously, and Dembia (12°23'26.2"N, 37°20'17.2"E) district, where chickpeas had been grown for several years.

2.3. Soil analysis and determination of most probable number of rhizobia (MPN)

Composite soil samples from both districts were collected from ten randomly selected sites from a 1ha area at the depths of 0-20 cm. Each composited soil sample was dried and passed through a 2 mm sieve. The soil physicochemical properties were analyzed following standard laboratory procedures in the National Soil Testing Center (NSTC), Addis Ababa. The rhizobia population from the soil was determined using the most-probable-number (MPN) plant infection method according to Howieson and Dilworth (2016), by inoculating soil dilutions on the host. Ten grams of soil was diluted in aseptic condition in 90 ml of sterilized distilled water. Then a serial dilution was performed, reaching a final dilution of 10^{-10} and used to inoculate a chickpea seedling adequately grown in acid-treated and sterilized sand using pots in three replications. The MPN was calculated from the number of units testing positive for each dilution to determine the factor (f). This factor (f) found in MPN tables then multiplied by the lowest dilution before all units were negative for nodulation.

The MPN was calculated from the most likely number (m) in MPN tables.

Number of rhizobia = $f \times d$, f − factor from the table, d −lowest dilution before all units were negative.

Table 1. The eco-physiological, IAR, PGP and symbiotic effectiveness of the selected rhizobia strains.

Isolates	**GUCR-19**	**GUCR-30**	**GUCR-55**
Relative species	*M. amorphae* strain B 19	*M.* sp. strain HKG 230	*M.* sp. strain UFLAO1-919
Accession number	PP529519	PP529520	PP529592
Temperature	15 - 45	20 - 45	15 - 45
Salt Max NaCl (%)	5	4	4
pH	6 - 10	5 - 10	5 - 10
Nitrogen source Utilization %	100	100	86
Carbon source Utilization %	100	100	100
IAR %	87.5	100	100
Phosphate solubilization (SI)	1.25	1.17	1.19
IAA production (μg/ml)	19.2	30.6	18.3
Antagonistic activity	-	+	+
Symbiotic Effectiveness (%)	82	91	84

Legend: IAR - Intrinsic antibiotic resistance; GUCR - Gondar University Chickpea rhizobia; IAA - Indole acetic acid, SE - Symbiotic effectiveness

Table 2. Stress tolerance, Plant growth promotion and biocontrol properties of the selected PGPR isolates.

Isolates	**GUCRB4**	**GUCRB21**	**GUCRB76**	**GUCRB124**
Relative species	*Alcaligenes* sp.	*Enterobacter mori*	*Serratia marcescens*	*Brevibacillus brevis*
Accession number	PP499249	PP499248	PP508218	PP499250
Temperature	20 - 50	20 - 50	20 - 50	10 - 45
Salt Max NaCl (%)	6	7	6	6
pH	5 - 9	6 - 10	5 - 9	5 - 10
IAR %	100	100	100	80
Phosphate solubilization (SI)	3.49	3.45	3.25	3.11
IAA production (μg/ml)	51.80	54.53	50.10	48.20
Ammonia Production	+	+	+	+
Cellulase Activity	+	+	+	+
Chitinase Production	+	+	+	+
Protease Activity	+	+	+	+
Siderophore Production	+	+	+	+
HCN Production	+	+	+	+
Antagonistic activity	+	+	+	+

Legend: GUCRB - Gondar University Chickpea Rhizobacteria; HCN - Hydrogen cyanide; IAA - Indole acetic acid; IAR - Intrinsic antibiotic resistance.

2.4. Inoculant preparation

The rhizobia and PGPR isolates selected from the stock culture were streaked on yeast extract mannitol agar (YEMA) and nutrient agar (NA) media, respectively and incubated at 28±2 °C for 3-5 days. A single colony from each isolate was inoculated into 100 ml yeast extract mannitol broth (YEMB) and nutrient broth (NB) media for rhizobia and PGPR respectively. The flasks were incubated at 28±2°C for 5-7 days on an incubator shaker at 120 rpm until it attained 10^9 cfu/ml.

2.5. Evaluating the compatibility of the isolates

The compatibility of the PGPR and rhizobia isolates was undertaken using cross streaking method on YEMA medium according to Martins *et al.* (2004).

2.6. Treatments and experimental design

The experiment was conducted at Bahirdar University, Department of Plant Science greenhouse. The composite soil was properly mixed, sieved and filled in 3 kg capacity surface sterilized plastic pots. Seeds of chickpea variety called 'Natoli' obtained from Debre Zeit Center, Ethiopian Institute of Agricultural Research (EIAR), were surface sterilized using 95% ethanol and 3% hypochlorite and rinsed in five changes of distilled sterilized water (Somasagaren and Hoben, 1994). Before planting, the soils were watered to approximately 75% field capacity. Five seeds of uniform size were sown in each pot and later thinned down to three after one week of germination. Each seedling was inoculated and co-inoculated with 1 ml (10^9 cfu/ml) of each isolate.

The microbial treatments included three single inoculations, co-inoculation, consortia and two controls. The experiment was arranged in a completely randomized design (CRD) with 18 treatments in triplicates (as described below). The pots were fertilized as per recommended by Somasegaran and Hoben (1994). The experiment consisted of uninoculated but nitrogen-fertilized pots as positive control and uninoculated non-nitrogen-fertilized pots as negative controls.

T1=GUCR-19 (*M. amorphae*)

T2=GUCR-30 (M. sp. HKG 230)

T3=GUCR-55 (M. sp. UFLAO1-919)

T4=GUCR-19 (*M. amorphae*) + GUCRB4 (*Alcaligenes* sp.)

T5=GUCR-19 (*M. amorphae*) + GUCRB21 (*Enterobacter mori*)

T6=GUCR-19 (*M. amorphae*) + GUCRB76 (*Serratia marcescens*)

T7=GUCR-19 (*M. amorphae*) + GUCRB124 (*Brevibacillus brevis*)

T8=GUCR-30 (*M.* sp. HKG 230) + GUCRB4 (*Alcaligenes* sp.)

T9=GUCR-30 (*M.* sp. HKG 230) + GUCRB21 (*E. mori*)

T10=GUCR30 (*M.* sp. HKG 230) + GUCRB76 (*S. marcescens*)

T11=GUCR-30 (*M.*sp. HKG 230) + GUCRB124 (*Brevibacillus brevis*)

T12=GUCR-55 (*M.*sp.UFLAO1-919) + GUCRB4 (*Alcaligenes* sp.)

T13=GUCR-55 (*M.* sp. UFLAO1-919) + GUCRB21 (*E. mori*)

T14=GUCR-55 (*M.* sp. UFLAO1-919) + GUCRB76 (*S. marcescens*)
T15=GUCR-55 (M. sp. UFLAO1-919) + GUCRB124 (*B. brevis*)
T16=Consortium [GUCR-19 (*M. amorphae*) + GUCR-30 (*M.* sp. HKG 230) + GUCR-55 (*M.* sp. UFLAO1-919) + GUCRB4 (A*lcaligenes* sp.) + GUCRB21 (*E. mori*) + GUCRB76 (*S. marcescens*) + GUCRB124 (*B. brevis*)]
T17=non-inoculated and unfertilized control
T18=non-inoculated but fertilized control

The pots were watered every two days and harvested after 8 weeks of planting to record the number of nodules (NN), nodule dry weight (NDW) and shoot dry weight (SDW). The number of nodules was counted as the mean number of nodules per plant. The nodule dry weight and shoot dry weight per plant were determined by oven drying the samples to constant weight at 70 °C for 48 hours. Dried shoot samples were finely ground to determine nitrogen content in (%) using the Kjeldahl method (Sahlemedhin and Taye, 2000). Symbiotic effectiveness (SE %) of each isolate was calculated according to Date (1993).

$$SE = \frac{\text{Shoot dry weight of plants inoculated with test isolate}}{\text{Shoot dry weight of plants supplied with nitrogen}} \times 100$$

The isolates were categorized by their symbiotic effectiveness as ineffective, <35%; lowly-effective, 35-50%; effective, 50-80%; and highly effective, >80%.

3. STATISTICAL ANALYSIS

The data were analyzed and interpreted using ANOVA. The experimental treatment means were compared and contrasted against their control and with each other following Duncan's multiple range test (DMRT) at a significance level of 0.05 using SPSS v. 25.

4. RESULTS

The effect of single and co-inoculation of rhizobia and PGPR isolates on the symbiotic effectiveness of chickpea was studied in pot experiments under greenhouse conditions, using two different soils (Table 3). The soil of the Dembia district is categorized as Vertisols on which chickpea had grown for several years and the soil of the Adet district is classified as Nitisols and had no previous history of inoculation. As indicated in Table 3, Dembia and Adet soils exhibited relatively neutral pH. The Dembia and Adet soils contained compatible native rhizobia of chickpea (Table 3), making them suitable for evaluating the competitive abilities of the rhizobia isolates against native field soil rhizobia. The Dembia soils with a long cropping history of chickpea harbored a relatively higher number of rhizobia nodulating chickpea. A relatively lower number of rhizobia was estimated in Adet soil which had no history of cropping chickpea.

The symbiotic effectiveness of single and co-inoculation with selected rhizobia (GUCR-19, 30 and 55) and PGPR (GUCRB-4, 21, 76 and 124) isolates on two different soils were presented in Tables 4 and 5.

Table 3. Chemical properties of the study soil

Parameters	Adet soil	Dembia soil
pH	6.86	7.18
EC (dS m^{-1})	0.003	0.13
CEC (cmol (+) kg^{-1})	32.6	41.6
Total nitrogen (%)	0.078	0.11
Organic carbon (%)	0.62	0.65
Available phosphorous (ppm)	1.34	2.9
Na (cmol (+) kg^{-1})	0.28	0.39
K (cmol (+) kg^{-1})	1.1	1.23
Fe (ppm)	17.7	16.9
Mn (ppm)	20.28	17.8
Zn (ppm)	0.34	0.46
Cu (ppm)	0.76	2.41
MPN	3×10^1	1×10^2

Table 4. Symbiotic effectiveness of the rhizobia and PGPR isolates on Adet soil

Treatments	NN/p	NDW (mg)	SDW (g)	SE (%)	TN (%)
GUCR-19 (*M. amorphae)*	36.8±1.0i	166.11±4.59f	2.600±0.025i	89	2.51
GUCR-30 (*M.* sp. HKG 230)	42.3±2.1hi	189.00±11.97d-f	2.868±0.039fg	98	2.70
GUCR-55 (*M.* sp. UFLAO1-919)	35.6±0.8i	158.78±2.73ef	2.696±0.043hi	92	2.48
GUCR-19 (*M. amorphae)* + GUCRB4 (*Alcaligenes* sp.)	46.9±1.6f-h	198.11±7.34c-e	2.836±0.049f-h	97	2.83
GUCR-19 (*M. amorphae)* + GUCRB21 (*Enterobacter mori*)	55.0±0.5de	207.66±2.11b-e	2.903±0.008e-g	99	2.89
GUCR-19 (*M. amorphae)* + GUCRB76 (*Serratia marcescens*)	63.6±3.3c	216.22±1.68a-c	3.021±0.030de	104	2.93
GUCR-19 (*M. amorphae)* + GUCRB124 (*Brevibacillus brevis*)	51.2±0.8e-g	193.44±11.12c-f	2.787±0.045f-h	96	2.79
GUCR-30 (*M. sp.* HKG 230) + GUCRB4 (*Alcaligenes* sp.)	53.0±2.0ef	192.56±1.89c-f	3.118±0.014cd	107	3.07
GUCR-30 (*M.* sp. HKG 230) + GUCRB21 (*Enterobacter mori*)	74.7±3.1b	226.33±3.92ab	3.440±0.025b	118	3.16
GUCR-30 (*M.* sp. HKG 230) + GUCRB76 (*Serratia marcescens*)	64.3±0.8c	205.56±13.51b-e	3.216±0.025c	110	3.27
GUCR-30 (*M.* sp. HKG 230) + GUCRB124 (*Brevibacillus brevis*)	55.2±1.7de	194.89±10.00c-f	3.132±0.028cd	107	3.18
GUCR-55 (*M.* sp. UFLAO1-919) + GUCRB4 (*Alcaligenes* sp.)	44.4±2.5gh	194.00±10.82c-f	2.757±0.024gh	95	2.75
GUCR-55 (*M.* sp. UFLAO1-919) + GUCRB21 (*Enterobacter mori*)	60.9±1.9cd	204.33±19.46b-e	2.923±0.043ef	101	2.84
GUCR-55 (*M.* sp. UFLAO1-919) + GUCRB76 (*Serratia marcescens*)	53.2±3.6ef	211.55±5.15b-d	2.816±0.091f-h	97	2.80
GUCR-55(*M.* sp. UFLAO1-919) + GUCRB124 (*Brevibacillus brevis*)	47.8±1.9e-h	186.67±5.09d-f	2.783±0.018f-h	96	2.76
Consortium	91.0±6.5a	237.00±3.51a	3.636±0.024a	125	3.90
Non-inoculated and unfertilized control (-VE)	11.7±1.0k	77.33±0.66h	1.688±0.065k		1.67
Non-inoculated but fertilized control (+VE)	5.3±0.3k	41.34±1.74i	2.907±0.011ef		2.73

Legend: Means followed by the same letter within a column are not significantly different at $P<0.05$ level, and all the data are the means of triplicates. NN/P - nodule number per plant, NDW - nodule dry weight, SDW - shoot dry weight, TN (%) - total nitrogen, SE - symbiotic effectiveness.

Nodulation of chickpea

In single inoculation, the three rhizobia isolates induced nodule numbers ranging from 37-46 /plant in Dembia soil and 35-42/plant in Adet soil. GUCR-30 (*M.* sp. HKG 230) scored the highest nodule number in both soils (Tables 4 and 5). Regarding the co-inoculation, the nodule number was in the range of 44-111/p showing that co-inoculation induced more nodule number than single inoculation. The highest nodule number (111/p and 91/p) was recorded by plants co-inoculated with the consortium in Dembia and Adet soil respectively, followed by 84/plant from plants co-inoculated with GUCR-30 (*M.* sp. HKG 230) + GUCRB21 (*Enterobacter mori*) in Dembia soil.

Table 5. Symbiotic effectiveness of the rhizobia and PGPR isolates on Dembia soil.

Treatments	NN/p	NDW (mg)	SDW (g)	SE (%)	T (
GUCR-19 (*M. amorphae)*	41.0±2.6hi	170.55±7.50h	2.839±0.021ij	101	2.
GUCR-30 (*M.* sp. HKG 230)	46.9±1.4g-i	195.11±9.98fg	3.116±0.034f-h	111	3.
GUCR-55 (*M.* sp. UFLAO1-919)	37.6±1.5i	173.67±3.38gh	2.777±0.068j	99	2.
GUCR-19 (*M. amorphae)* + GUCRB4 (*Alcaligenes* sp.)	58.4±0.8d-f	197.33±4.47ef	3.011±0.050hi	107	3
GUCR-19 (*M. amorphae)* + GUCRB21 (*Enterobacter mori*)	67.8±6.1cd	224.33±3.84c	3.303±0.059de	117	3
GUCR-19 (*M. amorphae)* + GUCRB76 (*Serratia marcescens*)	55.2±1.9e-g	202.33±10.39d-f	3.536±0.005c	126	3
GUCR-19 (*M. amorphae)* + GUCRB124 (*Brevibacillus brevis*)	49.7±1.9f-h	203.67±0.57d-f	3.214±0.053ef	114	3
GUCR-30 (*M.* sp. HKG 230) + GUCRB4 (*Alcaligenes* sp.)	59.1±0.4d-f	207.89±6.78c-f	3.263±0.012d-f	116	3
GUCR-30 (*M.* sp. HKG 230) + GUCRB21 (*Enterobacter mori*)	84.7±2.7b	260.00±1.73b	3.817±0.043b	136	3
GUCR-30 (*M.* sp. HKG 230) + GUCRB76 (*Serratia marcescens*)	68.4±2.8cd	222.00±7.79c	3.518±0.009c	125	3
GUCR-30 (*M.* sp. HKG 230) + GUCRB124 (*Brevibacillus brevis*)	63.8±3.9de	214.24±3.11c-e	3.431±0.031cd	122	3
GUCR-55 (*M.* sp. UFLAO1-919) + GUCRB4 (*Alcaligenes* sp.)	51.4±2.2f-h	191.33±5.56f-h	2.967±0.058h-j	105	3
GUCR-55 (*M.* sp. UFLAO1-919) + GUCRB21 (*Enterobacter mori*)	67.6±3.7cd	208.78±6.89c-f	3.197±0.104e-g	114	3
GUCR-55 (*M.* sp. UFLAO1-919) + GUCRB76 (*Serratia marcescens*)	76.3±6.4bc	219.22±1.17cd	3.432±0.038cd	122	3
GUCR-55(*M.* sp. UFLAO1-919) + GUCRB124 (*Brevibacillus brevis*)	53.0±1.7f-h	194.67±6.99fg	3.033±0.039gh	108	3
Consortium	111.3±6.6a	301.00±2.08a	4.323±0.129a	154	4
Non-inoculated and unfertilized control (-VE)	14.3±0.3k	92.33±0.66j	1.821±0.0301		1
Non-inoculated but fertilized control (+VE)	9.6±0.5 l	56.47 ±1.84 k	2.810±0.051 j		3

Legend: Means followed by the same letter within a column are not significantly different at P<0.05 level, and all the data are the means of triplicates. NN/P - nodule number per plant, NDW - nodule dry weight, SDW - shoot dry weight, TN (%) - total nitrogen, SE - symbiotic effectiveness.

Nodule dry weight

In this study, the nodule dry weight was between 158-195 mg/p in both soils in a single inoculation. The highest nodule dry weight 301 and 237 mg/p was accumulated by plants inoculated with consortium on Dembia and Adet soils respectively. In the co-inoculation following the consortium, the nodule dry weight was between 186-260mg/p (Tables 4 and 5). The highest (260 mg/p) was recorded from plants inoculated with GUCR-30 (*M.* sp. HKG 230) + GUCRB21 (*Enterobacter mori*) in Dembia soil, the lowest (186mg/p) was recorded from plants inoculated with GUCR-55 (*M.* sp. UFLAO1-919) + GUCRB124 (*Brevibacillus brevis*) in Adet soil.

Shoot dry weight

Regarding the shoot dry weight, in Dembia soil the highest shoot dry weight (4.323 g/p) was accumulated by plants inoculated with consortium followed by 3.817 g/p and 3.536 g/p co-inoculated with GUCR-30 (*M.* sp. HKG 230) + GUCRB21 (*Enterobacter mori*) and GUCR-19 (*M. amorphae)* + GUCRB76 (*Serratia marcescens*) which was 54%, 36% and 26% over the positive control respectively. In Adet soil the highest shoot dry weight (3.636 g/p) was accumulated by plants co-inoculated with consortium followed by 3.44 g/p and 3.216 g/p co-inoculated with GUCR-30 (*M.* sp. HKG 230) + GUCRB21 (*Enterobacter mori*) and GUCR-30 (*M.* sp. HKG 230) + GUCRB76 (*Serratia marcescens*), which was 25%, 19% and 11% over the positive control respectively (Tables 4 and 5).

Shoot total nitrogen

Regarding the shoot nitrogen content, chickpea inoculated with GUCR-30 (*M.* sp. HKG 230) + GUCRB21 (*Enterobacter mori*) (3.93%) in Dembia soil and GUCR-30 (*M.* sp. HKG 230) + GUCRB76 (*Serratia marcescens*) (3.27%) in Adet soil displayed the highest shoot nitrogen content (Tables 4 and 5).

Symbiotic effectiveness

The consortium inoculation showed the highest symbiotic effectiveness in both soils (154% in Dembia and 125% in Adet soil). In Dembia soil the three rhizobia isolates showed symbiotic effectiveness ranging from 99-111% and the co-inoculated plants showed symbiotic effectiveness ranging 105-136%. Chickpea plants co-inoculated with GUCR-30 (*M.* sp. HKG 230) + GUCRB21 (*Enterobacter mori*) showed the highest (136%) symbiotic effectiveness followed by GUCR-19 (*M. amorphae*) + GUCRB76 (*Serratia marcescens*) (126%). In Adet soil, the three rhizobia isolates showed symbiotic effectiveness ranging from 89-99% and the co-inoculated plants showed symbiotic effectiveness ranging 95-118%. Chickpea plants co-inoculated with GUCR-30 (*M.* sp. HKG 230) + GUCRB21 (*Enterobacter mori*) showed the highest (118%) symbiotic effectiveness followed by GUCR-30 (*M.* sp. HKG 230) + GUCRB76 (*Serratia marcescens*) (110%).

5. DISCUSSION

This study result showed that single and co-inoculation improved the nodulation, nodule dry weight and shoot dry weight of chickpea on both Dembia and Adet soils compared to the uninoculated negative controls (Figure 1). This indicates that the isolates have the potential to improve the growth of chickpea by different mechanisms such as increasing nutrient availability and uptake, producing plant hormones and suppressing soil-borne diseases.

In this study, single inoculation and co-inoculation induced a nodule difference of 23NN/p and 70NN/p over the negative control respectively which is relatively higher than Mirza *et al.* (2007) who reported that chickpea inoculated and co-inoculated with *Rhizobium* alone and *Rhizobium* Rnl + *Enterobacter* B showed a nodule number difference of 19 NN/p and 36 NN/p over the negative control, respectively. In single and co-inoculation the highest nodule number was scored in Dembia soil. This could be attributed to the difference in the indigenous rhizobia population which is higher in Dembia soil than Adet soil.

Co-inoculation treatments generally showed an average increase in nodule numbers by 21-125% compared to single inoculation treatments in both soils. This indicates that there was a synergistic interaction between rhizobia and PGPR isolates. However, Laabas *et al.* (2017) reported that mixed inoculation with selected rhizobial and PGPR strains was applied on two native soils, presenting a weak nodulation for the first and a total absence of nodulation for the second. Korir *et al.* (2017) reported that co-inoculation of the rhizobia strains with the PGPR generally enhanced the nodulation of common bean compared to single rhizobial inoculation.

Figure 1. Chickpea plant growth in the two soil types.

Adal (2018) reported an increase in nodule number by grass pea plants when inoculated with rhizobia and PGPR and consortium of the isolates than the single inoculation respectively. Similarly, Sintie (2018) reported that co-inoculation of rhizobia and PSRB isolates showed an almost a 64% increment in nodule number compared to their respective single inoculation of white lupin plant under pot soil culture. Different researchers also reported on the enhancement of nodulation by PGPR as they create more infection sites on the root systems of legume plants (Verma *et al.,* 2010; Badawi *et al.,* 2011; El-Nahrawy & Omara, 2017).

Regarding nodule dry weight, co-inoculation treatments showed an average increase in nodule dry weight up to 53% compared to single inoculation treatments in both soils. In this study, all single rhizobia and rhizobia + PGPR co-inoculated plants showed a nodule dry weight increment ranging from 78-168mg/p over the negative control in both soils. Mirza *et al.* (2007) reported nodule dry weight difference of 30mg/p and 20mg/p accumulated by chickpea plants co-inoculated and inoculated with *Rhizobium* Rnl + *Enterobacter* B and *Rhizobium* Rr2, respectively over the negative control.

In this study, co-inoculation treatments showed an average increment in shoot dry weight up to 25% compared to single inoculation treatments in both soils. Korir *et al.* (2017) reported that co-inoculation of rhizobia strains and PGPR recorded a higher shoot dry weight than single rhizobia inoculation in common beans.

In Dembia soil chickpea inoculated with GUCR-30 (*M.* sp. HKG 230) + GUCRB21 (*Enterobacter mori*) displayed the highest (3.93%) shoot nitrogen which showed 105%, 28% and 28-52% increment over the negative control, positive control and single inoculations. In Adet soil chickpea inoculated with GUCR-30 (M. sp. HKG 230) + GUCRB76 (Serratia marcescens) displayed the highest (3.27%) shoot nitrogen which showed 20 and 21-32% increment over the positive control and single inoculation respectively. This variation could be associated with the different adaptation abilities of the isolates to the soil environment. Many other studies have also shown positive effects of inoculation with PGPR strains on the growth of chickpea under greenhouse and field conditions (Valverd *et al.*, 2007; Malik & Sindhu 2011; Verma *et al.,* 2014) which could promote plant growth through different mechanisms such as the production of IAA, phosphates solubilization, siderophore production, and other plant-growth-promoting activities.

The symbiotic effectiveness result of this study indicated that single inoculated, co-inoculated and consortium treatments were highly effective in their symbiotic effectiveness.

In this study, among the inoculated, co-inoculated and control treatments, plants co-inoculated with the consortium showed the highest records in all measured parameters, next to the consortium plants co-inoculated with GUCR-30 (*M.* sp. HKG 230) + GUCRB21 (*Enterobacter mori*), GUCR-30 (*M.* sp. HKG 230) + GUCRB76 (*Serratia marcescens*) and GUCR-19 (*M. amorphae)* + GUCRB21 (*Enterobacter mori*), GUCR-19 (*M. amorphae)* +

GUCRB76 (*Serratia marcescens*), showed the highest in all parameters measured.

6. CONCLUSION

Co-inoculation of rhizobia and PGPR isolates led to a significant increase in nodule number, nodule dry weights, shoot dry weight and shoot total nitrogen compared to single inoculations, positive and negative controls. The isolates that showed the highest in all parameters measured can be used as inoculants to improve the growth of chickpea in soils that have similar soil characteristics as Dembia and Adet districts. In this study, chickpea planted in Dembia soil showed a better performance in all parameters than in Adet soil. A further field experiment in different locations is recommended for upgrading these isolates into chickpea inoculants.

7. REFERENCES

Abd El-Azeem SAM. (2022). Effects of co-Inoculation of Rhizobium and plant growth-promoting Rhizobacteria on common bean (*Phaseolus vulgaris*) Yield, Nodulation, Nutrient Uptake, and Microbial Activity under Field Conditions. *J. Soil Water Sci.*, 7(1): 13-26.

Adal M. (2018). Diversity, symbiotic and plant growth promoting properties of rhizobia and rhizobacteria of grass pea (*Lathyrus sativus* L.) from Central Ethiopia: implication to the selection and use of microbial inoculants in low input agriculture in Ethiopia. PhD dissertation, Addis Ababa University, Addis Ababa, Ethiopia.

Aslam M, Jiang C, Zafar U, Usama M, Haroon H. (2018). Functional genomics prospective of Chickpea breeding: Constraints and future directions. *Modified Conceptual Development Agronomy*, 3(4): 327-332.

Asrat Mulat, Giziew Almaz, and Ayalew Dereje (2022). "Analysis of Actors in Chickpea Seed Supply Chain; Gondar Zuria Woreda, Ethiopia." *Ethiopian Journal of Agricultural Sciences.* 32(2): 99–117.

Assefa Funga, Chris, O. Ojiewo, Losenge Turoop, Githiri Stephen Mwangi (2016). Symbiotic effetiveness of elite rhizobia strains nodulating Desi type chickpea (*Cicer arietinum* L.) varieties. *J. Pl Sci,* 4: 88–94.

Assefa Gonfa, Saba Fetene and Sosena Amsalu (2024). Validation and Demonstration of Improved Soil and Crop Management Practices for Chickpea (*Cicer arietinum* L.) Productions in the Central Highlands Vertisols of Ethiopia. *Journal of Soil Science & Plant Physiology*

Atsede Muleta, Kassahun Tesfaye, Teklehaimanot Haileselassie, Douglas R. Cook and Fassil Assefa (2021). Screening for Ecologically Competent, Nutritional Characteristics and Symbiotically Effective Chickpea Nodulating *Mesorhizobium spp.* Isolated from Acidic Soils of Ethiopia. *Journal of Microbial & Biochemical Technology*. 13(8):1-11.

Badawi FSF, Biomy AMM, Desoky AH. (2011). Peanut plant growth and yield as influenced by co-inoculation with *Bradyrhizobium* and some rhizo-microorganisms under sandy loam soil conditions. *Ann. Agric. Sci.,* 56(1): 17-25.

Beneduzi A, Ambrosini A, Passaglia LM. (2012). Plant growth-promoting rhizobacteria (PGPR): their potential as antagonists and biocontrol agents. *Genet. Mol. Biol.,* 35:1044-1051.

Bharti N, Pandey SS, Barnawal D, Patel VK, Kalra A. (2016). Plant growth promoting rhizobacteria *Dietzia natronolimnaea* modulates the expression of stress responsive genes providing protection of wheat from salinity stress. *Sci. Rep.,* 6(1): 1-16.

CSA (Central Statistical Agency) (2022). Agricultural sample survey, area and production of major crops (Private Peasant Holdings, Meher Season). Addis Ababa, Ethiopia.

Denton MD, Pearce DJ, Peoples MB. (2013). Nitrogen contributions from faba bean (*Vicia faba* L.) reliant on soil rhizobia or inoculation. *Plant Soil*, 365: 363-374.

El-Nahrawy S, Omara AED. (2017). Effectiveness of co-inoculation with *Pseudomonas koreensis* and rhizobia on growth, nodulation and yield of common bean (*Phaseolus vulgaris* L.). *Microbiol. Res. J. Int.,* 21: 1-16.

Fahde S, Boughribil S, Sijilmassi B, Amri A. (2023). Rhizobia: a promising source of plant growth-promoting molecules and their non-legume interactions: examining applications and mechanisms. *Agriculture,* 13(7): 1279.

FAOSTAT (2023). United Nation's Food and Agricultural Organization. Statistical Division. Rome.

FAOSTAT (2024). United Nation's Food and Agricultural Organization. Statistical Division. Rome.

Fasusi OA, Cruz C, Babalola OO. (2021). Agricultural sustainability: microbial biofertilizers in rhizosphere management. *Agriculture*, 11(2): 163.

Gashaw Nahusenay, Girma Wolde, Wondwosen Tena, Tatek Tamiru (2024). Chickpea (*Cicer arietinum L.*) growth, nodulation, and yield as affected by varieties, *Mesorhizobium* strains, and NPSB fertilizer in Southern Ethiopia. *Front. Plant Sci.* 15:1372082.

Glick BR, Gamalero E. (2021). Recent developments in the study of plant microbiomes. *Microorganisms,* 9(7): 1533.

Hawi Negusse, Teklehaimanot Haileselassie, Mulatu Geleta and Kassahun Tesfaye (2025). Genetic Variability of Ethiopian Chickpea (*Cicer arietinum* L.) Landraces for Acid Soil Tolerance. *Plants*, 14(3): 311.

Howieson JG, Dilworth MJ. (2016). Working with rhizobia. Canberra: Australian Centre for International Agricultural Research, p. 173.

Hungria M, Nogueira MA, Araujo RS. (2013). Co-inoculation of soybeans and common beans with rhizobia and azospirilla: strategies to improve sustainability. *Biol. Fertil. Soils.*, 49: 791-801.

ICRISAT (2023). ICRISAT Annual Report: Commemorating 50 Years of Impact. International Crops Research Institute for the Semi-Arid Tropics.

Ilyas N, Mumtaz K, Akhtar N., Yasmin, H., Sayyed, R.Z., Khan, W., Enshasy, H.A.E., Dailin, DJ, Elsayed EA, Ali Z. (2020). Exopolysaccharides producing bacteria for the amelioration of drought stress in wheat. *Sustainability,* 12(21): 8876. *J. Curr. Microbiol. Appl. Sci.,* 9:136-151.

Jang JH, Woo SY, Kim SH, Khaine I, Kwak MJ, Lee HK, Lee TY, Lee WY. (2017). Effects of increased soil fertility and plant growth-promoting rhizobacteria inoculation on biomass yield, energy value, and physiological response of poplar in short-rotation coppices in a reclaimed tideland: A case study in the Saemangeum area of Korea. *Biomass & Bioenergy*, 107: 29-38.

Jiménez-Díaz, R. M., Castillo, P., Jimenez-Gasco, M. del M., Landa, B. B., & Navas-Cortes, J. A. (2015). Fusarium wilt of chickpeas: Biology, ecology and management. *Crop Protection*. Advance online publication. https://doi.org/10.1016/j.cropro.2015.02.023

Korir H, Mungai NW, Thuita M, Hamba Y, Masso C. (2017). Co-inoculation effect of rhizobia and plant growth promoting rhizobacteria on common bean growth in a low phosphorus soil. *Frontiers in plant science*, 8: 141.

Kumar A, Maurya BR, Raghuwanshi R. (2021). The microbial consortium of indigenous rhizobacteria improving plant health, yield and nutrient content in wheat (*Triticum aestivum*). *J. Plant Nutr*., 44(13): 1942-1956.

Kumari P, Sharma P, Sharma S. (2020). Synergism of rhizobium and rhizobacteria on growth, symbiotic parameters, soil quality and grain yield in summer mungbean (*Vigna radiata* l. wilczek). *Int.*

Laabas S, Boukhatem ZF, Bouchiba Z, Benkritly S, Abed NE, Yahiaoui H, Bekki A, Tsaki H. (2017). Impact of single and co-inoculations with Rhizobial and PGPR isolates on chickpea (*Cicer arietinum* L.) in cereal-growing zone soil. *J. Plant Nutr.,* 40(11): 1616-1626.

Malik DK, Sindhu SS. (2011). Production of indole acetic acid by Pseudomonas sp.: effect of coinoculation with *Mesorhizobium* sp. Cicer on nodulation and plant growth of chickpea (*Cicer arietinum* L.). *Physiol. Mol. Biol. Plants*., 17: 25-32.

Martins A, Kimura O, Goi SR, Baldani JI. (2004). Effect of the co-inoculation of plantgrowth promoting rhizobacteria and rhizobia on development of common bean plants (*Phaseolus vulgaris* L.). Floresta e Ambiente, 11(2): 33-39.

Midekssa MJ, Löscher CR, Schmitz RA, Assefa F. (2016). Phosphate solubilization and multiple plant growth promoting properties of rhizobacteria isolated from chickpea (*Cicer aeritinum* L.) producing areas of Ethiopia. *Afr. J. Biotechnol.,* 15(35): 1899-1912.

Mirza BS, Mirza MS, Bano A, Malik KA. (2007). Coinoculation of chickpea with Rhizobium isolates from roots and nodules and phytohormone-producing *Enterobacter* strains. *Aust. J. Exp. Agric*., 47(8): 1008-1015.

MOA (2023). Transforming Ethiopian Agriculture: Power Point Presentation, Briefing for Agricultural Scholar Consultative Forum, April 2023, Addis Ababa.

Nigam, S N and Chaudhari, S and Deevi, K C and Saxena, K B and Janila, P. (2021). Trends in Legume Production and Future Outlook. In: *Genetic Enhancement in Major Food Legumes*. pp. 7-48 Springer, Cham, Switzerland

Pandey P, Bisht S, Sood A, Aeron A, Sharma GD, Maheshwari DK. (2012). Consortium of plant-growth-promoting bacteria: future perspective in agriculture. Bacteria in agrobiology: *plant probiotics*, 185-200.

Rajendran G, Sing F, Desai AJ, Archana G. (2008). Enhanced growth and nodulation of pigeon pea by co-inoculation of *Bacillus* strains with *Rhizobium* spp. *Bioresour. Technol.*, 99(11): 4544-4550.

Sahlemedhin S., Taye B. (2000). Procedures for soil and plant analysis. Tech. Pap., 74: 110.

Sánchez AC, Gutiérrez RT, Santana RC, Urrutia AR, Fauvart M, Michiels J, Vanderleyden J. (2014). Effects of co-inoculation of native *Rhizobium* and *Pseudomonas* strains on growth parameters and yield of two contrasting *Phaseolus vulgaris* L. genotypes under Cuban soil conditions. *Eur. J. Soil Biol.*, 62: 105-112.

Santoyo G, Guzmán-Guzmán P, Parra-Cota FI, Santos-Villalobos SDL, Orozco-Mosqueda MDC, Glick BR. (2021). Plant growth stimulation by microbial consortia. *Agronomy,* 11(2): 219.

Setotaw Ferede, Nigusie Girma, Sherif Aliy (2019). Validating product concepts for opti-mizing chickpea breeding. 122, 61.

Shugute Addisu, Taye Tessema and Takele Negewo (2023). Distribution and importance of weeds in wheat, and farmers' practices to manage weeds at Chelia and Midakegn Districts, West Shoa Zone, Oromiya, Ethiopia. *Global Journal of Agriculture & Veterinary Science.* 1(1): 5–11.

Sintie E. (2018). Diversity, symbiotic effectiveness and plant growth promoting characteristics of Rhizobia and Rhizospheric Bacteria on growth and production of White lupin (*Lupinus albus* L.) under greenhouse and field conditions in North Western Ethiopia. PhD dessertation, Addis Ababa University.

Sofi, S.A.; Muzaffar, K.; Ashraf, S.; Gupta, I.; Mir, S.A. (2020). Chickpea. In *Pulses*; Manickavasagan, A., Thirunathan, P., Eds.; pp. 55–76 Springer International Publishing: Cham, Switzerland, ISBN 978-3-030-41376-7.

Somasegaran P., Hoben HJ. (1994). Handbook for rhizobia: methods in legume-Rhizobium technology. Springer Science & Business Media.

Tafesse Gezahegn, Kifle Tesfamariam, Tekeste Berhanu, Miet Maertens, Steven Van Passel, Marijke D'Haese (2020). Conflict, fraud, and distrust in Ethiopian agricultural cooperatives. *Journal of Co-operative Organization and Management*. 8. 100106.

Tahir HA, Gu Q, Wu H, Raza W, Hanif A, Wu L, Colman MV, Gao X. (2017). Plant growth promotion by volatile organic compounds produced by Bacillus subtilis SYST2. *Front. Microbiol.,* 8: 171.

Tariku Tirfeessa, Adeba Gemechu and Lama Zemedu (2025). The impact of climate variability on chickpea production in the central highlands of Ethiopia using auto regressive distributive lag model. *Discov Agric.* 3: 73

Tibebu Belete, Firew Mekibib, Million Eshete (2017). Genetic Improvement of Desi Type Chickpea (*Cicer arietinum* L.) Varieties Yield Potential and Its Agronomic Trait Determinants in Ethiopia from 1974-2010. *Journal of Biology, Agriculture and Healthcare*. 7: 84-99.

Ullah N, Ditta A, Khalid A, Mehmood S, Rizwan MS, Ashraf M, Mubeen F, Imtiaz M, Iqbal MM. (2020). Integrated effect of algal biochar and plant growth promoting rhizobacteria on physiology and growth of maize under deficit irrigations. *J. Soil Sci. Plant Nutr.*, 20: 346-356.

Valverde A, Burgos A, Fiscella T, Rivas R, Velazquez E, Rodríguez-Barrueco C, Cervantes E, Chamber M., Igual JM. (2007). Differential effects of coinoculations with *Pseudomonas jessenii* PS06 (a phosphate-solubilizing bacterium) and *Mesorhizobium ciceri* C-2/2 strains on the growth and seed yield of chickpea under greenhouse and field conditions. – In: First international meeting on microbial phosphate solubilization, Springer Netherlands, p. 43-50.

Verma JP, Yadav J, Tiwari KN, Jaiswal DK. (2014). Evaluation of plant growth promoting activities of microbial strains and their effect on growth and yield of chickpea (*Cicer arietinum* L.) in India. *Soil Biol Biochem.,* 70: 33-37.

Verma JP, Yadav J, Tiwari KN, Lavakush S, Singh V. (2010). Impact of plant growth promoting rhizobacteria on crop production. *Int. J. Agric. Res.*, *5*(11):

954-983.

Verma P, Yadav AN, Khannam KS, Kumar S, Saxena AK, Suman A. (2016). Molecular diversity and multifarious plant growth promoting attributes of Bacilli associated with wheat (*Triticum aestivum* L.) rhizosphere from six diverse agro-ecological zones of India. *J. Basic Microbiol.,* 56(1): 44-58.

Yilma Abebe, Merkuz Abera, Berhanu Bekele and Tesfaye Alemu (2025). Prevalence, pathogen diversity and management practices of chickpea Fusarium wilt in Ethiopia. *CABI Reviews*, 20(1), p.0043

Zehara Mohammed, Cook, D., Greenlon, A., Asnake Fikre, Wettberg, E., Marques, E. (2020). Ecological Competence, Plant Growth Promoting and Symbiotic Characteristics of Different *Mesorhizobium* Strains Nodulating Chickpea (*Cicer arietinum* L.) from Ethiopia. *J Plant Pathol Microbiol*. 11: 1-11.

Zewdie A. (2018). Assessment of chickpea seed borne disease with special reference to Ascochyta blight (*Didymellarabiei*) in Central Ethiopia. *Int. J. Life Sci.,* 6(3): 707-712.

Zhang J, Cook J, Nearing JT, Zhang J, Raudonis R, Glick BR, Langille MG, Cheng Z. (2021). Harnessing the plant microbiome to promote the growth of agricultural crops. *Microbiol. Res*., 245, 126690.

Publisher: Eliva Press Global Ltd

Email: info@elivapress.com

www.ingramcontent.com/pod-product-compliance
Lightning Source LLC
LaVergne TN
LVHW022325200926
848481LV00004B/1433
* 9 7 8 9 9 9 9 3 4 5 0 3 3 *